Word Problems 2

Len and Anne Frobisher

Heinemann
Halley Court, Jordan Hill, Oxford, OX2 8EJ
a division of Reed Educational and Professional Publishing Ltd

Heinemann is a registered trademark of Harcourt Education Limited

ISBN 0 435 20862 4 (Pupils' book)
ISBN 0 435 20863 2 (Teacher's version)

09 08 07 06 05 04 03 02

Illustrated by Garry Davies

Typesetting and layout by Bookcraft Ltd, Stroud, Gloucestershire

Printed by Ebenezer Baylis, Worcester

Contents

Number problems 1

count in 1s / 10s / 100s, + facts to 10, addition doubles

1 A number is falling off.

What number is falling off? 8

2 Jack has a tube of 6 Sunbursts and a tube of 3 Sunbursts.

How many Sunbursts does Jack have altogether? 9

3 There are 5 fish in each tank.

How many fish are in the two tanks? 10

4 Mia has 4 packs of crayons.

How many crayons does Mia have? 40

5 Max counts in hundreds.

What number should Max say next? 500

6 The bus numbers go down in tens.

What will be the number on the last bus? 30

7

I am 2 more than 5.

What am I? 7

Number problems 2

1/10 more/less, numbers in figures/words, + facts to 10, halves

1 A supermarket has 47 check-outs. One is closed.

How many are open? 46

2 At Hopton School 73 children are ill.

Write 73 in words. Seventy-three

3 There are 63 people waiting for a ride. Ten more join the line.

How many are there now? 73

4 There are 7 balloons at a party. Pat bursts 4 of them.

How many are left? 3

5 Lucy has 18 toy cars.

Ben has half as many as Lucy. How many cars has Ben? 9

6 Nine children like red apples best. Two children like green apples best.

How many more like red apples than green ones? 7

7 *I am 10 less than 17.*

What am I? 7

Number problems 3

partition,
+/– 9/11,
10× facts, doubles

1 Harry bakes 27 buns. He puts two lots of 10 into boxes.

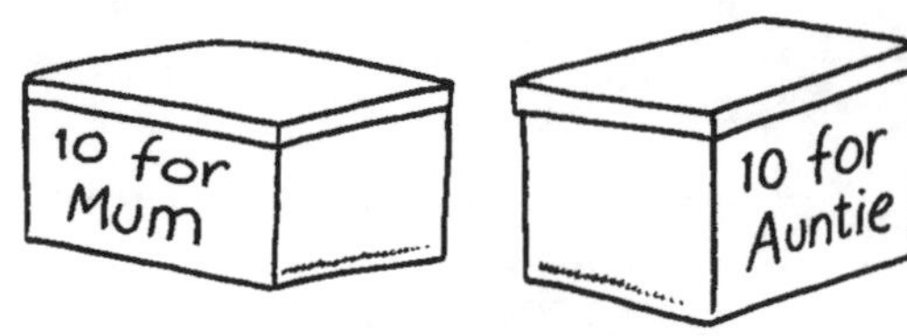

How many buns are left? 7

2 There are 36 people on a bus. Then 9 more get on.

How many are on the bus now? 45

3 There are 53 children in the pool. Eleven of them are boys.

How many are girls? 42

4 There are 10 biscuits in a pack.

Altogether how many biscuits are in 7 packs? 70

5 Sam has 7 cards. Tom has double that.

How many cards has Tom? 14

6 A stall has 46 story books. It sells 9 of them.

How many are left? 37

I am in the 10 times table.
I am between 65 and 75.

What am I? 70

Money problems 1

totals and change, near doubles, 10× facts, +/− facts to 10

1 Bill has £2. His dad gives him £1. His mum gives him 50p.

How much does Bill have now? £3.50

2 A spaceship costs £8. A rocket costs £7.

How much does it cost to buy both of them? £15

3 Zoe buys some chews. They cost 3p each.

What is the cost of 10 chews? 30p

4 A pencil costs 7p. A rubber costs 2p.

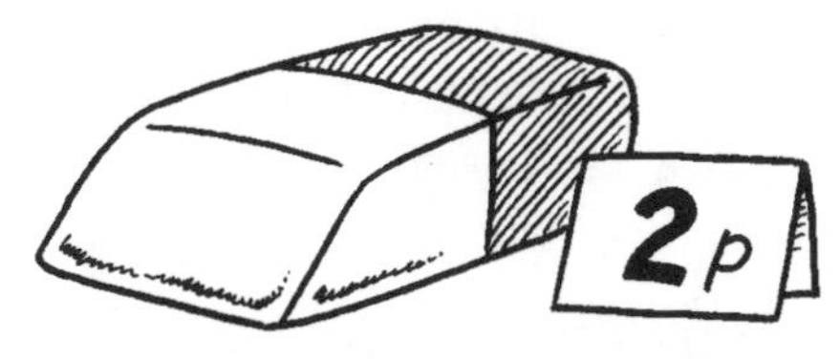

How much does Stacey need to buy both? 9p

5 Millie has £4. She buys a comic for £1.

How much has Millie left? £3

6 In one pocket Ajit has a 5p coin. In his other pocket he has three 1p coins.

He spends 6p.
How much has he left? 2p

7

I am a silver coin. I am more than 10p and less than 50p.

What am I? 20p

Length problems 1

compare, suggest units, +/– facts to 10, doubles

1 Greg makes a kite that is 45 cm long. Rod makes one that is 40 cm long.

Who makes the longer kite? Greg
How much longer is it? 5 cm

2 Abbie hops 2 metres then jumps 4 metres.

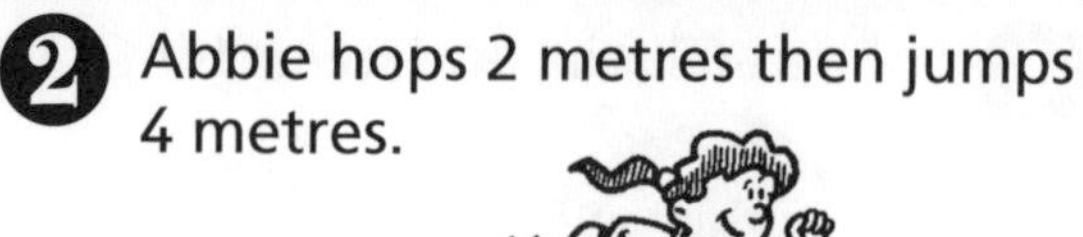

Altogether how far does Abbie hop and jump? 6 m

3 Tariq has a rubber that is 3 cm long. Lucy's rubber is twice as long as Tariq's.

How long is Lucy's rubber? 6 cm

4 James' carpet is 6 metres long. He cuts some off so it is 5 metres long.

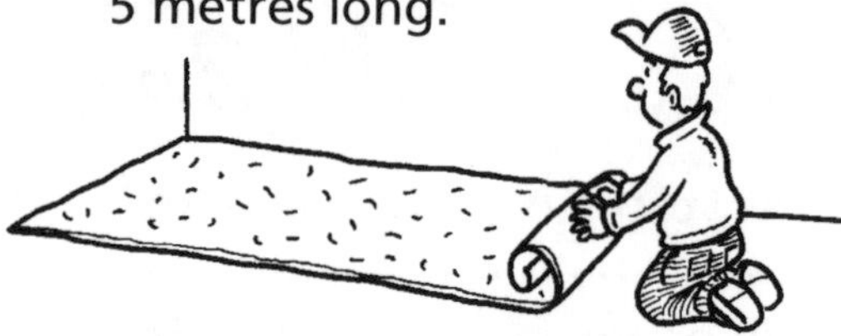

What length does he cut off? 1 m

5 Liza is measuring the width of her table.

What unit of length do you think she should use? centimetres

6 Luke's tower is 10 cm tall.
He knocks 3 cm off the top.
Then he adds some more blocks to make it 2 cm taller.

How tall is the tower now? 9 cm

7

I am double one metre.

What am I? 2 m

Time problems 1

units of time, 10× facts, +/− facts to 10, halves

1 Lee leaves school at 3 o'clock.

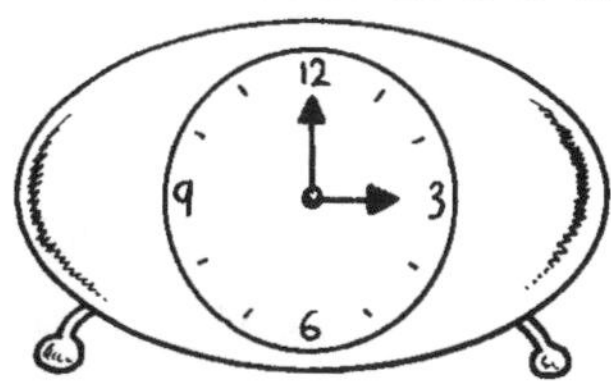

He takes 15 minutes to walk home. At what time does Lee get home? 3:15

2 Owen goes on a ferry. He sleeps for 4 hours of the journey and is awake for 5 hours.

How long is the journey? 9 hrs

3 Alex balances a ball on his head for 18 seconds. Liam does it for only half this time.

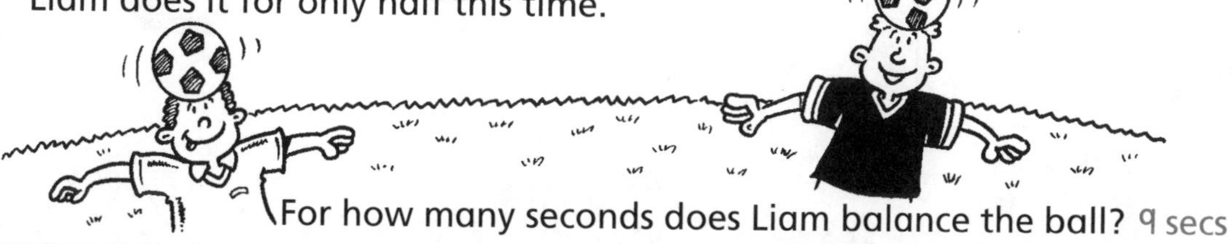

For how many seconds does Liam balance the ball? 9 secs

4 Ten children take turns to read a book for charity. Each child reads for 5 minutes.

What is the total time for the ten children? 50 mins

5 Rob and Rachel stand on one leg. Rob does it for 2 minutes, Rachel does it for 5 minutes.

How much longer does Rachel stand on one leg than Rob? 3 mins

6 Emily and Sam go on holiday. They spend one day with their aunt then 7 days with their grandparents. It rains on 4 days.

On how many days of the holiday does it not rain? 4 days

7

I am a whole number of hours. I am between half-past 10 and half-past 11.

What am I? 11 o'clock

Review problems 1

1. How many children are at the bonfire? 15
How many grown-ups are there? 12
Altogether how many people are at the bonfire? 27

2. How many sparklers are in each box? 10
How many boxes of sparklers are on the table? 4
How many sparklers are there altogether? 40

3. At the start there were 10 toffee apples for sale.
How many are left on the stall? 6
How many have been sold? 4

4. Mrs Patel buys 7 candy floss with a £1 coin.
What do they cost? 70p
How much change does she get? 30p

5. Here are 2 rockets.
Which rocket is longer? Shooter
How many centimetres longer? 5 cm

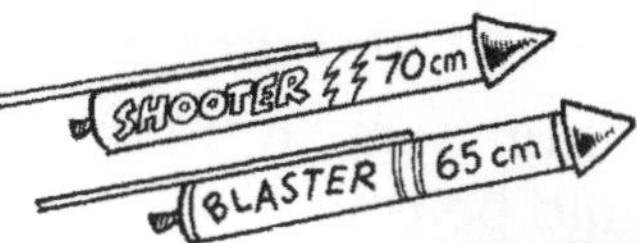

6. The Shooter rocket goes 10 metres high.
The Blaster rocket goes to half this height.
How high does the Blaster go? 5 m

7. The bonfire starts at 6 o'clock.
It will end at 8 o'clock.
How long will the bonfire last? 2 hrs

8. Mrs Hall brought 11 fireworks.
Mr Roy brought 13 fireworks.
How many did they bring altogether? 24

9. The Magic Fountain shoots stars 2 metres high.
How many centimetres is that? 200 cm

10. The Vulcan shoots stars 10 cm higher than the Magic Fountain.
How high do its stars go? 210 cm

Number problems 4

count on/back in 2s and 10s, odd/even numbers

1 The number on Jade's house has fallen off.

What should the number be? 23

2 One of the balls has an odd number.

What is the odd number? 7

3 The seat numbers go down in 10s.

What is the hidden seat number? 13

4 The numbers on the hooks should be odd.

Which number is not correct? 8
What should the number be? 7

5 The chalets are numbered with even numbers.

What is the number of the chalet behind the tree? 14

6 Amelia counts in 10s starting at 4.

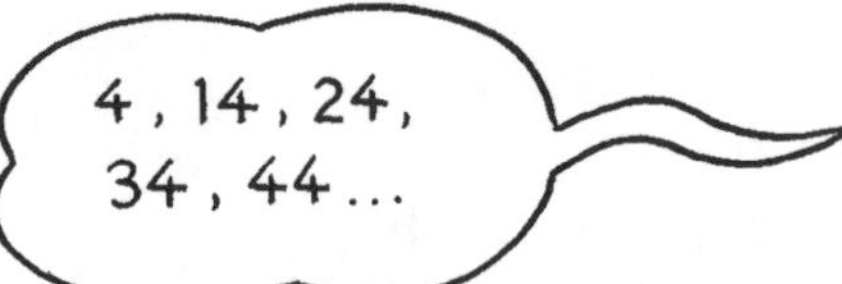

What number should Amelia say next? 54

7

I am three 10s more than 21.

What am I? 51

Number problems 5

compare, order, ordinal numbers, small difference, +/− facts to 10

1 Dulip has 52 picture cards.
Ryan has 55 cards.

Who has more cards? Ryan
How many more? 3

2 Sammy has 41 books on her shelf and Tyrone has 38 on his shelf.

Who has more books? Sammy
How many more? 3

3 On a boat trip Jamie sees 2 whales. Later he sees another 8 whales.

How many whales does he see altogether? 10

4 For a school trip children are put into 10 groups. Each group has 6 children.

How many children go on the trip? 60

5 Five children line up in order. Two of them are in the wrong places.

Which two are in the wrong places? 44 and 46

6 These are the riders in a race.

John is third.
What is his number? 2

7

I am half-way between 10 and 20.

What am I? 15

Money problems 2

£.p notation, count in 100s, doubles, pairs that total 20, 10× facts

1 Irfan collects pennies.
He puts them in boxes of 100.

How many pence has he in his 4 boxes? 400
How much is this in £s? £4

2 Molly's dad says, 'If you save £12, I will double it.'

How much will Molly have after her dad has doubled her £12? £24

3 Kieran buys a 14p choc bar and a 6p apple.

What is the total cost of the choc bar and the apple? 20p

4 Toy penguins cost £4 each. Ashley's mum buys 10 to put in a parade.

How much do the 10 penguins cost? £40

5 Callum buys the stickers with a 20p coin.

How much change does Callum get? 7p

6 Lily is given five £1 coins for her birthday.

How many pence is her money worth? 500p

7

I am the smallest silver coin.

What am I? 5p

Number problems 6

repeated addition,
× facts,
recognise halves,
+/– 11

1 There are 25 sheep in a field. Eleven more sheep are put in the field.

How many sheep are in the field now? 36

2

How many wheels on 1 trike? 3

How many wheels on 5 trikes? 15

3 Ella thinks each person at her party will eat 2 biscuits.

There will be 6 people.

How many biscuits does she need to make? 12

4 In a wood there were 34 trees. Eleven of the trees were blown over in a storm.

How many were left standing? 23

5 Hasan eats half a bar of chocolate. Hannah eats the other half.

How much of the chocolate is left? none

6 Alex builds 7 sandcastles.
Megan builds twice as many as Alex.

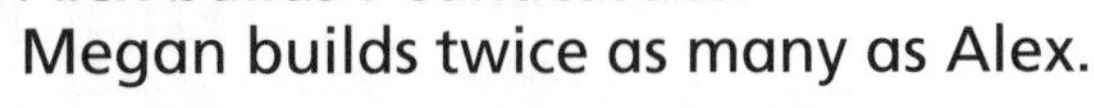

How many does she build? 14

7

I am an odd two-digit number. My digits add to make 3.

What am I? 21

Length problems 2

interpret tables, near doubles, +/− 9, nearest cm

1 A pile of computer games is 12 cm high. Another pile is 13 cm high.

How high would the games be if they were in one pile? 25 cm

2 A railway engine and its tender are 15 metres long. The engine is 9 metres long.

How long is the tender? 6 m

3 This table shows the heights of four children.

name	height
Will	119 cm
Erin	120 cm
Reece	116 cm
Ella	122 cm

Who is the tallest? Ella

Who is the shortest? Reece

4 A line of boys is 15 metres long. A line of girls is 16 metres long.

If the girls and boys make one line, how long will it be? 31 m

5 Nicole's left foot is just over 17½ cm long.

What is the length of Nicole's left foot to the nearest cm? 18 cm

6 A hall is 18 metres long. The dining room is 9 metres long.

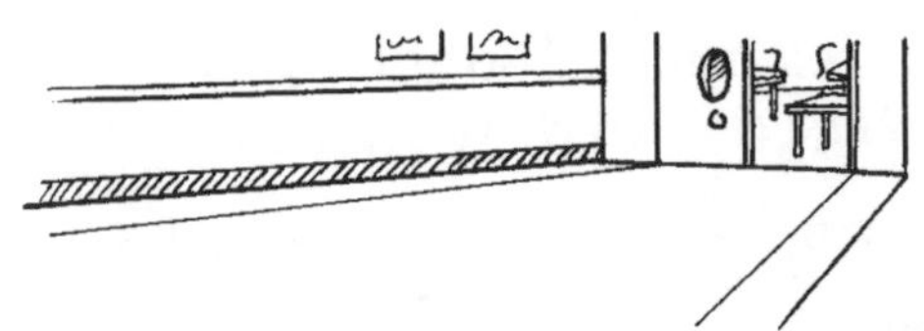

What is the total length of the hall and the dining room? 27 m

7

I am a length in cm. Both my digits are even and they add to make 4.

What am I? 22 cm

Time problems 2

interpret tables,
order months,
read time,
halves, ÷ by 10

1 The table shows the birthdays of five children.
Whose birthday is last every year? Leah
Whose birthday comes third every year? Lauren

name	birthdays
Sam	30 March
Leah	11 December
Lauren	4 July
Ethan	22 May
Haroon	8 October

2 Shannon looks at the clock.
In half an hour she must leave for school.

At what time must Shannon leave for school? 8:30 or half-past 8

3 Connor writes his full name in 20 seconds.
Amber writes hers in half the time.

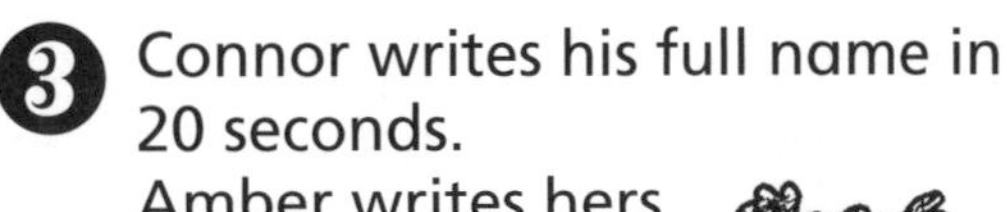

How long does it take for Amber to write her name? 10 secs

4 Ten people cycle for 30 hours in a relay race. Each one cycles for the same length of time.

For how long does each person cycle? 3 hrs

5 Charlie looks at the clock.
In 15 minutes it is lunchtime.

What time is lunch? 12:15

6 The clock shows the end of silent reading.
It lasted for half an hour.

At what time did silent reading start? 10:30 or half-past 10

7

I am a time in the morning. My number of hours belongs to the 10 times table. My number of minutes is half of 60.

What am I?
10:30

Review problems 2

review of previous content

1 In Year 1 there are 48 children.
There are 52 in Year 2.

How many children are there altogether in Years 1 and 2? 100

2 Bethany has a £20 note.
She buys an £11 jigsaw.

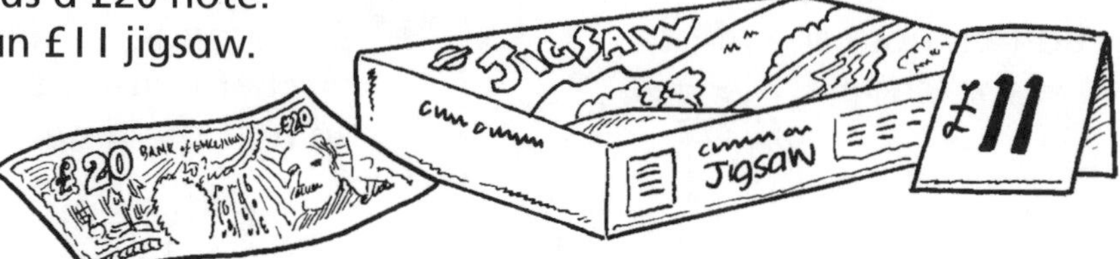

How much change does she get? £9

3 Bradley's baby sister is 47 cm when born.
She grows 9 cm in the next 3 months.

How long is she then? 56 cm

4 Salman gets home from school at 4 o'clock. It takes him half an hour.

At what time does he leave school? 3:30 or half-past 3

5 There are 3 ducks on a pond. Five more ducks fly in and 2 ducks fly away.

How many ducks are then on the pond? 6

6 Molly has saved £24. She is given another £23 for her birthday.

How much has she now? £47

7

When 11 is added to me the answer is 20.

What am I? 9

Number problems 7

recognise multiples of 5, count objects in 2s/5s/10s, +/– facts to 10, doubles

1 There are 2 buns in each box.

How many buns are in 6 boxes? 12

2 In a game Jake scores 8 points.

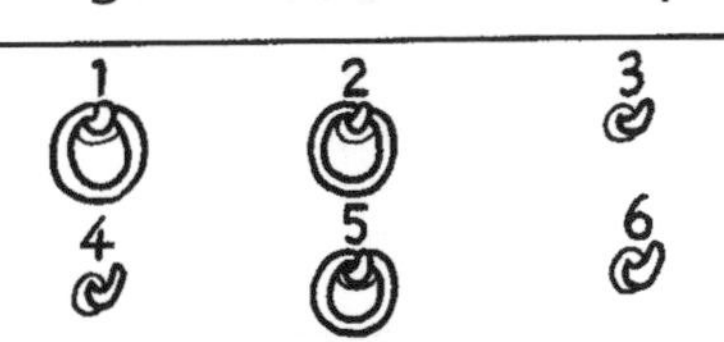

Julia scores twice as many points as Jake.
How many points does Julia score? 16

3 On each sheet there are 10 stamps.

How many stamps are on 3 sheets? 30

4 Ethan counts up in 3s.

Which is the first multiple of 5 that Ethan says? 15 or fifteen

5 Jade cuts each cake into 5 pieces.

How many pieces are there altogether? 20

6 In a swimming pool there are 10 children.
Seven of them get out and 3 more get in.

How many are in the pool now? 6

7

I am a multiple of 5.
The sum of my 2 digits is 8.

What am I? 35

Number problems 8

compare 2-digit numbers,
+ several numbers,
10× facts, halves

1 On a bus there are 6 passengers.
At a stop 2 more get on.
At the next stop 7 more get on.

How many are on the bus now? 15

2 There are 81 fish in Sid's tank.
In Lily's tank there are 69 fish.

Whose tank has more fish? Sid's
How many more? 12

3 At a bus station there are 10 buses. Half of the buses are double-deckers.

How many of the buses are double-deckers? 5

4 At Terminal 1 there are 47 planes.
At Terminal 2 there are 62 planes.

How many planes are there altogether? 109

5 On a day trip there are 16 people.
One-half of them are children.

How many are children? 8

6 A boat can carry 8 people.

A full boat makes 10 journeys.
How many people does it carry altogether? 80

7

I am more than 20 and less than 30.
The difference between my digits is 5.

What am I? 27

Money problems 3

totals and change, +/– facts to 20, pairs that total 20

1 Tad buys one toffee bar and one chew.

What is the total cost? 10p

2 Holly gave the shopkeeper a £10 note. She was given a £5 note as change.

How much did Holly spend at the shop? £5

3 Amin has one 10p coin and two 5p coins in his pocket.

How much has Amin in his pocket? 20p

4 A red T-shirt costs £6 and a white one £3.

How much more does the red T-shirt cost than the white one? £3

5 At the Summer Fair Grace buys an 8p bun and a 2p drink.

How much does she pay altogether? 10p

6 Bruce has a 20p coin.
He buys a tube of Fizzers for 10p.
Later he finds 5p that he had lost.

How much does he have now? 15p

7

I am the sum of one of each of the copper coins.

What am I? 3p

Number problems 9

partition to + 6/7/8/9,
2× facts,
near doubles

1 A bus has 8 people upstairs and 6 people downstairs.

How many people are on the bus? 14

2 Each car on the ride can carry 4 people.

How many people can 2 cars carry? 8

3 Daisy buys 2 bunches of flowers.
One bunch has 6 flowers, the other has 9 flowers.

How many flowers does Daisy buy? 15

4 Each netball team has 7 players and 1 reserve.

How many children are there altogether in two teams? 16

5 Zoe buys 2 bags of carrots.
One bag has 22 carrots,
the other has 23 carrots.

How many carrots does Zoe have altogether? 45

6 In a marina there are 9 boats.
7 more boats sail in.

How many are in the marina now? 16

7

I am an even number.
My tens digit is 6 less than my units digit.

What am I? 28

Time problems 3

relationship between units,
+9/11,
÷ by 10 facts

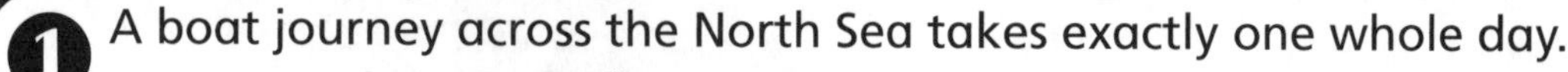

1 A boat journey across the North Sea takes exactly one whole day.

How many hours is the journey? 24 hrs

2 One Monday Sue says that in 9 days she will go on holiday.

On what day will she go on holiday? Wednesday

3 Class 2 have 1 hour in which to make a model castle.

How many minutes do they have? 60 mins

4 It took Sam 60 seconds to put on his shoes.

How long was that in minutes? 1 min

5 In 11 hours time the Smith family fly to Disneyland.
What time will the clock show when they fly? 6 o'clock or 6:00

6 Ten children have the same amount of time on a swing.
Altogether they spend 60 minutes on the swing.

How long is each child on the swing? 6 mins

7

I am the number of months in 2 years.

What am I? 24

Weight problems 1

reading scales,
– 9/11, ÷ by 2 facts,
count in 100s

1 A box of apples weighs 22 kg.
Then 9 kg of apples are sold.

What is the weight of apples left in the box? 13 kg

2 Sally weighs 5 kg of potatoes.
She puts another 2 kg of potatoes on the scales.

How many kilograms are on the scales? 7 kg

3 Tom buys 5 Choco bars.

What is total weight of the 5 bars? 500 g

4 Georgia weighs 48 g of flour.
She takes 11 g of flour off the scales.

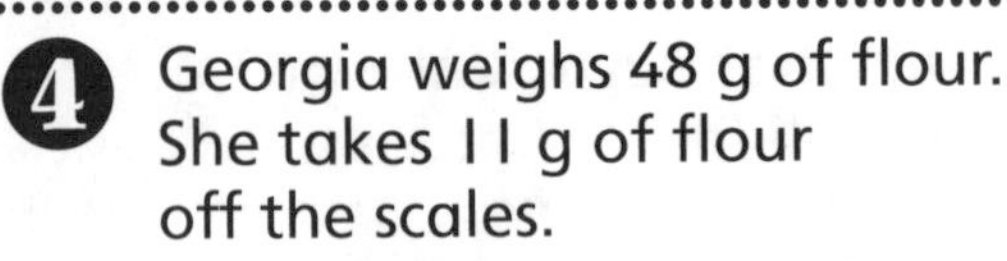

How many grams of flour are left? 37 g

5 A large bag of sugar weighs 8 kg.

How many 2 kg bags can be filled from the large bag? 4

6 Sophie buys 2 bags of pears.
One weighs 6 kg, the other 2 kg.

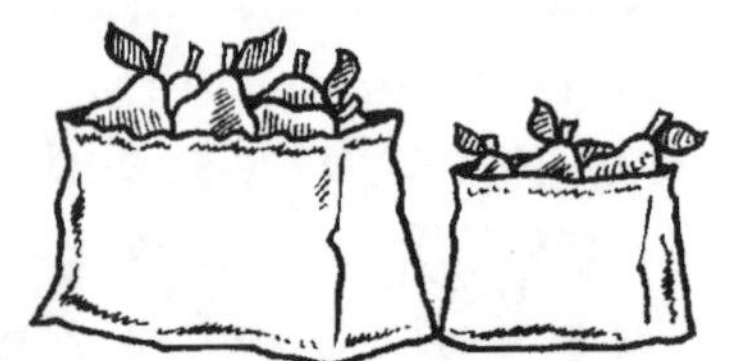

What is the difference in the weights of the bags? 4 kg

7

I am less than 80 kg and more than 70 kg. The difference between my digits is 5.

What am I?

72 kg

Teacher's Notes

Contents

Introduction

The pupils' books

Word Problems, a series of five books, one for each of the years 2 to 6, helps develop children's ability to solve number problems in a variety of contexts. Each page of word problems may be used to support numeracy lessons taught earlier in a week or for homework. The books provide weekly practice of word problems, as described in the National Numeracy Strategy (NNS) *Framework* and are designed to match the weekly structure within the NNS *Sample Medium-term Plans*. The style of questions reflects the examples that appear in the National Curriculum tests.

The books contain two types of page:

Topic pages: Each page has six word problems which are devoted solely to one topic (Number, Money, Time, Length, Weight, Capacity and Measures). The mathematical content of each page is listed on page iii. The mathematical content of the word problems on each page is listed in the lozenge at the top of the page.

Question 7 on each Topic page is a number puzzle, which gives children practice in Reasoning with Number using a variety of number properties. These questions are similar to those in the *Framework* and the National Curriculum tests.

Review pages look back at the mathematical content of previous Topic pages. Some Review pages are double page spreads, with the left page having a scene with information that is required to answer questions, and the right page asking questions about the scene.

Illustrations are used to tune children in to the 'real life' context of the word problem. Each is part of the problem and it is important that children look at the illustrations closely. An illustration may contain information that is in the word problem itself. On some occasions an illustration will contain information that is not in the text of the word problem, but is essential in order to solve the problem.

Answers to word problems are printed in red at the side of each question in the pupils' pages of the Teacher's notes.

Helping children solve word problems

Here are some suggestions for helping children develop a strategy for solving word problems.

- Make sure children read a question carefully and don't merely search for key words such as 'altogether' which they think, sometimes incorrectly, tell them what to do with the numbers. At the start of the year you may wish to read a question with children until they become independent readers.
- Encourage children to close their eyes and picture the context of a problem and any actions that are performed with/on 'objects' in the context.
- Allow them to talk with each other about a problem and to ask themselves:
  - 'What do I have to find out?'
  - 'What do I know that will help me find out?'
  - 'What do I have to do with what I know to find out?'

It is also important as part of a word problem-solving strategy that children develop:

- the skill of distinguishing information that is helpful from that which is not helpful
- the ability to choose and use appropriate operations to solve word problems.

These can only develop with experience of solving many word problems over a long period of time.

Teaching plans

So that you can integrate *Word Problems* into your medium-term teaching plans pages iv to vi show the relationship between the Topics in *Word Problems* and the sample medium-term plans suggested by the NNS. Pages vii and viii show how the topics in *Word Problems* relate to Mathematics in the National Curriculum in Wales Programme of Study and the similar Programme of Study for Northern Ireland.

Summary of mathematical content

Page	*Topic*	*Mathematical content*
4	Number 1	count in 1s/10s/100s; + facts to 10; addition doubles
5	Number 2	1/10 more/less; numbers in figures and words; + facts to 10; halves
6	Number 3	partition; +/– 9/11; 10x facts; doubles
7	Money 1	totals and change; near doubles; 10x facts; +/– facts to 10
8	Length 1	compare; suggest units; +/– facts to 10; doubles
9	Time 1	units of time; 10x facts; +/– facts to 10; halves
10–11	**Review 1**	**review of previous content**
12	Number 4	count on/back in 2s and 10s; odd/even numbers
13	Number 5	compare; order; ordinal numbers; small difference; +/– facts to 10
14	Money 2	£.p notation; count in 100s; doubles; pairs that total 20; 10x facts
15	Number 6	repeated addition; x facts; recognise halves; +/– 11
16	Length 2	interpret tables; near doubles; +/– 9; nearest cm
17	Time 2	interpret tables; order months; read time; halves; ÷ by 10
18	**Review 2**	**review of previous content**
19	Number 7	recognise multiples of 5; count objects in 2s/5s/10s; +/– facts to 10; doubles
20	Number 8	compare 2-digit numbers; + several numbers; 10x facts; halves
21	Money 3	totals and change; +/– facts to 20; pairs that total 20
22	Number 9	partition to + 6/7/8/9; 2x facts; near doubles
23	Time 3	relationship between units; + 9/11; ÷ by 10 facts
24	Weight 1	reading scales; – 9/11; ÷ by 2 facts; count in 100s
25	**Review 3**	**review of previous content**
26	Number 10	count in 5s; 10 more/less; 2x facts
27	Number 11	more/less; between; bridging through 10/20; + facts to 10; 10x facts
28	Money 4	totals and change; + 2-digit numbers; – facts to 10; ÷ by 10 facts
29	Number 12	sharing; grouping; facts/pv to ÷; halves; pairs that total 20
30	Money 5	½/¼ of objects; +/– facts to 10; ÷ by 2 facts
31	Measures 1	read half-hour; kilograms; doubles; pairs of multiples of 10 total 100
32–33	**Review 4**	**review of previous content**
34	Number 13	count in 3s/4s; 5x facts; +/– facts to 10
35	Number 14	order; round to nearest 10; doubles of multiples of 5; +/– facts to 10
36	Money 6	four operations; x/÷ by 2
37	Number 15	facts/pv to x/÷; 2x facts; pairs of multiples of 10 total 100
38	Capacity 1	read scales; suggest units; ÷ by 2 facts; 2x and 10x facts
39	Time 4	read half-hours; relationship between units; ÷ by 10 facts
40–41	**Review 5**	**review of previous content**
42	Number 16	count on/back in 3s/4s; +/– 9/11
43	Number 17	order; +/– less than 20; + single digit; 5x facts; pairs of multiples of 10 total 100
44	Money 7	+/– 19/21; 5x facts; double multiples of 5
45	Number 18	halving as inverse of doubling; facts/pv to x/÷; halve multiples of 10; ÷ by 10 facts
46	Money 8	totals and change; coins; 2/4=1/2; +/– facts to 10; x by 2
47	Measures 2	read half and quarter hours on clock; pairs that total 20; ÷ by 2 facts
48	**Review 6**	**review of previous content**

Word Problems and the NNS *Sample Medium-term Plans*

AUTUMN			
Sample medium-term plans		*Word Problems*	
Unit	*Topic*	*Pages*	*Topic*
1	Counting, properties of numbers and number sequences	4	Number 1
2–4	Place value, ordering, estimating, rounding Understanding + and – Mental calculation strategies (+ and –) Money and 'real life' problems Making decisions, checking results	5–7	Number 2 Number 3 Money 1
5–6	Measures, including problems Shape and space Reasoning about shapes	8–9	Length 1 Time 1
7	**Assess and review**	**10–11**	**Review 1**
8	Counting, properties of numbers and number sequences Reasoning about numbers	12	Number 4
9	Place value, ordering, estimating, rounding Understanding + and – Mental calculation strategies (+ and –) Money and 'real life' problems Making decisions, checking results	13–14	Number 5 Money 2
10–11	Understanding x and ÷ Mental calculation strategies (x and ÷) Money and 'real life' problems Making decisions, checking results Fractions	15	Number 6
12–13	Measures, and time, including problems Handling data	16–17	Length 2 Time 2
14	**Assess and review**	**18**	**Review 2**

SPRING			
Sample medium-term plans		*Word Problems*	
Unit	*Topic*	*Pages*	*Topic*
1	Counting, properties of numbers and number sequences	19	Number 7
2–4	Place value, ordering, estimating, rounding Understanding + and – Mental calculation strategies (+ and –) Money and 'real life' problems Making decisions, checking results	20–22	Number 8 Money 3 Number 9
5–6	Measures, including problems Shape and space Reasoning about shapes	23–24	Time 3 Weight 1
7	**Assess and review**	**25**	**Review 3**
8	Counting, properties of numbers and number sequences Reasoning about numbers	26	Number 10
9	Place value, ordering, estimating, rounding Understanding + and – Mental calculation strategies (+ and –) Money and 'real life' problems Making decisions, checking results	27 28	Number 11 Money 4
10	Understanding x and ÷ Mental calculation strategies (x and ÷) Money and 'real life' problems Making decisions, checking results Fractions	29–30	Number 12 Money 5
11–12	Measures, and time, including problems Handling data	31	Measures 1
13	**Assess and review**	**32–33**	**Review 4**

SUMMER			
Sample medium-term plans		*Word Problems*	
Unit	*Topic*	*Pages*	*Topic*
1	Counting, properties of numbers and number sequences	34	Number 13
2–4	Place value, ordering, estimating, rounding Understanding + and – Mental calculation strategies (+ and –) Money and 'real life' problems Making decisions, checking results	35–37	Number 14 Money 6 Number 15
5–6	Measures, including problems Shape and space Reasoning about shapes	38–39	Capacity 1 Time 4
7	**Assess and review**	**40–41**	**Review 5**
8	Counting, properties of numbers and number sequences Reasoning about numbers	42	Number 16
9	Place value, ordering, estimating, rounding Understanding + and – Mental calculation strategies (+ and –) Money and 'real life' problems Making decisions, checking results	43–44	Number 17 Money 7
10–11	Understanding x and ÷ Mental calculation strategies (x and ÷) Money and 'real life' problems Making decisions, checking results Fractions	45–46	Number 18 Money 8
12–13	Measures, and time, including problems Handling data	47	Measures 2
14	**Assess and review**	**48**	**Review 6**

Word Problems and the National Curriculum in Wales

Using and Applying Mathematics

U1. Making and Monitoring Decisions to Solve Problems
U2. Developing Mathematical Language and Communication
U3. Developing Mathematical Reasoning

Number

N1. Understanding Number and Place Value
N2. Understanding Number Relationships and Methods of Calculation
N3. Solving Numerical Problems
N4. Classifying, Representing and Interpreting Data

Shape, Space and Measures

S3. Understanding and Using Measures

Word Problems		*Relevant sections of the National Curriculum Programme of Study*							
Page	*Topic*	*U1*	*U2*	*U3*	*N1*	*N2*	*N3*	*N4*	*S3*
4	Number 1	x	x	x	x	x	x		
5	Number 2	x	x	x	x	x	x		
6	Number 3	x	x	x	x	x	x		
7	Money 1	x	x	x	x	x	x		
8	Length 1	x	x	x					x
9	Time 1	x	x	x					x
10–11	**Review 1**	x	x	x	x	x	x		x
12	Number 4	x	x	x	x	x	x		
13	Number 5	x	x	x	x	x	x		
14	Money 2	x	x	x	x	x	x		
15	Number 6	x	x	x	x	x	x		
16	Length 2	x	x	x				x	x
17	Time 2	x	x	x				x	x
18	**Review 2**	x	x	x	x	x	x		x
19	Number 7	x	x	x	x	x	x		
20	Number 8	x	x	x	x	x	x		
21	Money 3	x	x	x	x	x	x		
22	Number 9	x	x	x		x	x		
23	Time 3	x	x	x					x
24	Weight 1	x	x	x					x
25	**Review 3**	x	x	x	x	x	x		x
26	Number 10	x	x	x	x	x	x		
27	Number 11	x	x	x	x	x	x		
28	Money 4	x	x	x	x	x	x		
29	Number 12	x	x	x		x	x		
30	Money 5	x	x	x	x	x	x		
31	Measures 1	x	x	x					x
32–33	**Review 4**	x	x	x	x	x	x		x
34	Number 13	x	x	x	x	x	x		
35	Number 14	x	x	x	x	x	x		
36	Money 6	x	x	x		x	x		
37	Number 15	x	x	x		x	x		
38	Capacity 1	x	x	x					x
39	Time 4	x	x	x					x
40–41	**Review 5**	x	x	x	x	x	x	x	x
42	Number 16	x	x	x	x	x	x		
43	Number 17	x	x	x	x	x	x		
44	Money 7	x	x	x		x	x		
45	Number 18	x	x	x		x	x		
46	Money 8	x	x	x	x	x	x		
47	Measures 2	x	x	x					x
48	**Review 6**	x	x	x	x	x	x		x

Word Problems and the National Curriculum in Northern Ireland

PROCESSES IN MATHEMATICS

P1. Using Mathematics
P2. Communicating Mathematically
P3. Mathematical Reasoning

HANDLING DATA

H1. Collect, Represent and Interpret Data

NUMBER

N1. Understanding Number and Number Notation
N2. Patterns, Relationships, and Sequences
N3. Operations and their Applications
N4. Money

MEASURES (M)

Page	*Topic*	*P1*	*P2*	*P3*	*N1*	*N2*	*N3*	*N4*	*M*	*H1*
4	Number 1	x	x	x	x	x	x			
5	Number 2	x	x	x	x	x	x			
6	Number 3	x	x	x	x	x	x			
7	Money 1	x	x	x				x		
8	Length 1	x	x	x					x	
9	Time 1	x	x	x					x	
10–11	**Review 1**	x	x	x	x	x	x	x	x	
12	Number 4	x	x	x	x		x			
13	Number 5	x	x	x	x	x	x			
14	Money 2	x	x	x				x		
15	Number 6	x	x	x	x	x	x			
16	Length 2	x	x	x					x	x
17	Time 2	x	x	x					x	x
18	**Review 2**	x	x	x	x	x	x	x	x	
19	Number 7	x	x	x	x	x	x			
20	Number 8	x	x	x	x	x	x			
21	Money 3	x	x	x				x		
22	Number 9	x	x	x		x	x			
23	Time 3	x	x	x					x	
24	Weight 1	x	x	x					x	
25	**Review 3**	x	x	x	x	x	x	x	x	
26	Number 10	x	x	x	x	x	x			
27	Number 11	x	x	x	x	x	x			
28	Money 4	x	x	x				x		
29	Number 12	x	x	x		x	x			
30	Money 5	x	x	x				x		
31	Measures 1	x	x	x					x	
32–33	**Review 4**	x	x	x	x	x	x	x	x	
34	Number 13	x	x	x	x	x	x			
35	Number 14	x	x	x	x	x	x			
36	Money 6	x	x	x				x		
37	Number 15	x	x	x		x	x			
38	Capacity 1	x	x	x					x	
39	Time 4	x	x	x					x	
40–41	**Review 5**	x	x	x	x	x	x	x	x	x
42	Number 16	x	x	x	x	x	x			
43	Number 17	x	x	x	x	x	x			
44	Money 7	x	x	x				x		
45	Number 18	x	x	x		x	x			
46	Money 8	x	x	x				x		
47	Measures 2	x	x	x					x	
48	**Review 6**	x	x	x	x	x	x	x	x	

Review problems 3

review of previous content

Altogether how many paper clips are in the 10 boxes? 1000

2 Faheem leaves school at 3 o'clock. Seven hours later he flies on a plane to see his grandparents.

At what time does Faheem fly?
10 o'clock or 10:00

3 Mary pays for the magazine with a £2 coin.

How much change does she get? 50p

4 The Watson family of 11 adults and 9 children visit Sealife Centre.

How many are there in the Watson family altogether? 20

5 Pat has two money boxes. She writes labels to show how much they hold.

£1·30

How much has she altogether in the two boxes? £2.50

6 Leo weighs 17 kg.
Aziz weighs 21 kg.

What would the scales show if they stood on them together? 38 kg

I am four 100s more than 23.

What am I? 423

Number problems 10

count in 5s, 10 more / less, 2× facts

1 At lunch time there are 24 children in the playground. Ten more children join them.

How many are in the playground now? 34

2 Each pack of pens has 5 red ones and 5 blue ones.

How many red pens are in 3 packs? 15
How many blue pens are in 3 packs? 15
Altogether how many pens are in 3 packs? 30

3 There are 6 cards in each packet.

How many cards are in 2 packets? 12

4 In Class 2 there are 41 children.
In Class 1 there are 10 fewer children.

How many children are in Class 1? 31

5 Two groups of children have tennis lessons. In each lesson there are 9 children.

How many children have lessons? 18

6 Brett has 4 boxes of marbles. There are 5 marbles in each box.

Tony has 10 fewer marbles than Brett.
How many marbles has Tony? 10

7

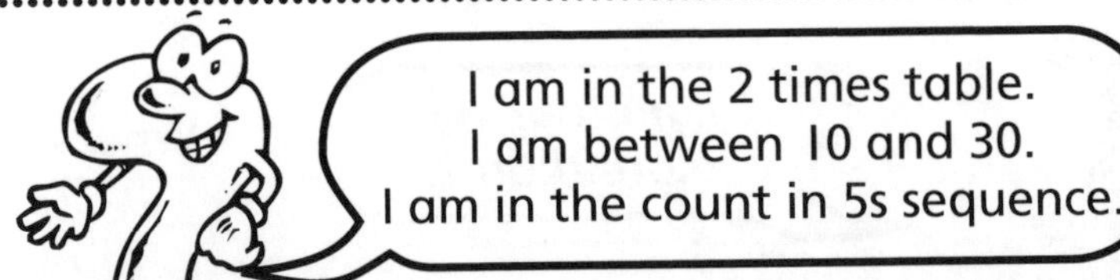

What am I? 20

Number problems 11

more/less, between, bridging through 10/20, + facts to 10, 10× facts

1 Sally saw 77 butterflies at a butterfly farm. Leon saw 53.

Who saw more? Sally
How many more? 24
Who saw fewer? Leon
How many fewer? 24

2 At the Spring Fair there were 2 rides for children over 6 years old and 6 rides for all ages.

How many rides were there altogether? 8

3 The numbers on the houses go up in ones. Some numbers have worn off.

Which numbers are missing? 22, 23, 24

4 There are 16 dogs each in its own kennel. Nineteen kennels are empty.

How many kennels are there altogether? 35

5 On sports day the children are put into 10 teams. There are 10 children in each team.

Altogether how many children are in the teams? 100

6 Danny puts out 34 biscuits on plates. Nineteen of them are eaten.

How many biscuits are left? 15

7

When you take 8 from me you get 12.

What am I? 20

Money problems 4

totals and change,
+ 2-digit numbers,
– facts to 10,
÷ by 10 facts

1 Meera has two purses.
In one purse she has 25p, in the other she has 32p.

How much does Meera have altogether? 57p

2 Beth has £7.
She buys a book for £6.

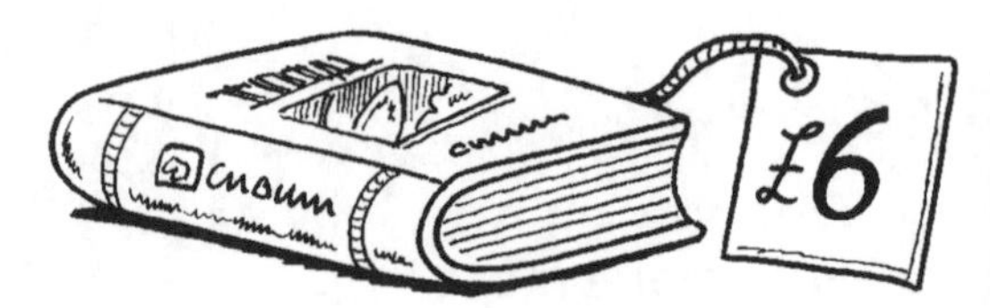

How much has Beth left? £1

3 Sean has 30p. He spends 23p.

How much change does he get? 7p

4 Ten dog biscuits cost 50p.

How much is one dog biscuit? 5p

5 Abdul has two £1 coins, one 50p coin and two 20p coins.

Which coins should he use to buy the comic? £1, 50p, 20p

6

What is the total cost of the 2 rubbers? 32p

7

I am the largest value coin.

What am I? £2

Number problems 12

sharing, grouping, facts/pv to ÷, halves, pairs that total 20

1 A book shop has 11 copies of *Magic for 7 Year Olds* on a shelf and 9 copies on display.

How many copies does it have altogether? 20

2 Sue and Stan share equally the 8 cherries in the bowl.

How many cherries do they each get? 4

3 Ella has 22 toy animals. She gives half of them to her little brother.

How many does she give to her brother? 11

4 Dad blows up 15 balloons. He gives each child 5 balloons.

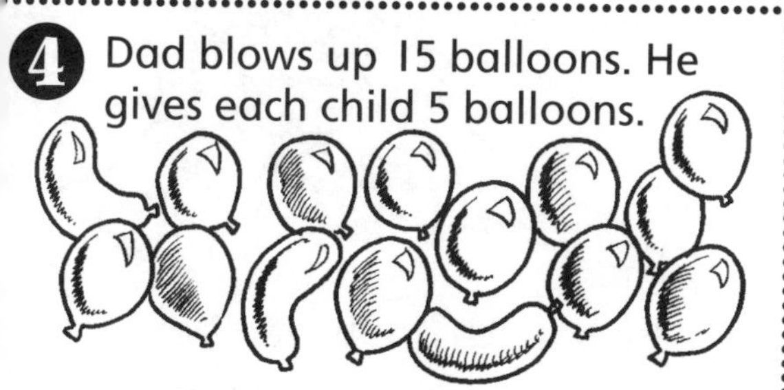

How many children get 5 balloons? 3

5 20 children visit a zoo. Fourteen of them want to see the snakes.

How many do not want to see the snakes? 6

6 The school has two buses to take 60 children on a visit. Each bus holds the same number of children.

How many children are on each bus? 30

7

I am 3 less than a half of 20. What am I? 7

Money problems 5

½/¼ of objects, +/– facts to 10, ÷ by 2 facts

1 Grace has 20p. She gives half of it to Darren.

How much has she left? 10p

2 Mum pays £10 for two packs of dice.
There are three dice in each pack.

What is the cost of one pack of dice? £5

3 Connor has 8p.

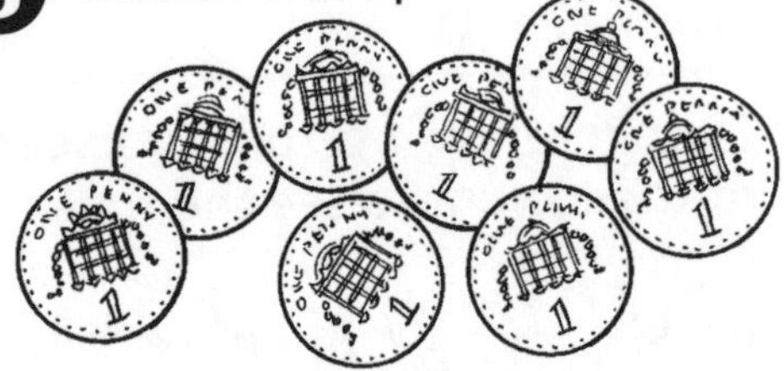

He spends a quarter of his 8p.
How much does he spend? 2p

4 Matt has £9. He spends £7 on some flowers for his mum's birthday.

How much has he left? £2

5 Jenny has 6p.
She loses half of it. Her mum then gives her another 4p.

How much has she now? 7p

6 Jordan has saved £4.

He gets £6 for his birthday and spends £5.
How much has he left? £5

7

I am the total of 2 different seven-sided coins.

What am I? 70p

Measures problems 1

read half-hour, kilograms, doubles, pairs of multiples of 10 total 100

1 A baker has 6 kg of flour. He needs double that amount.

How much flour does the baker need? 12 kg

2 Amber has two dogs. One weighs 40 kg, the other weighs 60 kg.

She weighs both dogs together. How many kilograms will the scales show? 100 kg

3 The station clock shows 8:30.

The train should leave in half an hour. At what time should it leave?
9:00 or 9 o'clock

4 On a stall are 10 kg of potatoes. The man sells 4 kg. Then he puts out 3 kg more on the stall.

How many kilograms of potatoes are on the stall now? 9 kg

5 Amy buys a 70 g packet and a 30 g packet of crisps.

What is the total weight of the two packets? 100 g

6 A plane is due to leave at 2:30. It is delayed half an hour.

At what time does the plane leave? 3:00 or 3 o'clock

7

I am an o'clock. Each of my 2 digits is a straight line.

What am I? 11 o'clock or 11:00

Review problems 4

review of previous content

1 Bilton School visit a farm.
They use 2 buses.
Each bus has 8 adults and 40 children.
How many adults are on the 2 buses? 16
How many children are on the 2 buses? 80

2 These are the coins Simon has in his purse.
How much does Simon have? 45p

3 The buses left school at 9:30.
The journey took 45 minutes.
At what time did they arrive at the farm? 10:15

9:30

4 How many children have got off the white bus at the farm? 4
How many children are still on the white bus? 36

5 The are 40 children on the grey bus.
Each of the 8 adults looks after the same number of children.
How many children are with each adult? 5

6 How many sheep and lambs are in the field? 9
Tomorrow the farmer will put in 10 more sheep and 16 more lambs.
How many animals will there be in the field then? 35

7 The black sheep weighs 71 kg. The white sheep weighs 66 kg.
Which sheep is heavier? the black sheep
How much heavier? 5 kg

8 Half of the piglets belong to each sow.
How many piglets does each sow have? 6

9 There are 20 chicks altogether at the farm. Six are outside and the rest are in their house.
How many chicks are not in their house? 14

10 Each tractor has 4 wheels and 1 spare wheel.
How many wheels are there altogether on the 3 tractors? 15

Number problems 13

count in 3s/4s,
5× facts,
+/– facts to 10

1 Tennis balls are in tubes of 3.

How many balls in 5 tubes? 15

2 A pet shop has cages with 2 gerbils in each.

How many gerbils are in 5 cages? 10

3 Brendan buys 2 packs of 3 ice-lollies and 2 packs of 4 ice-lollies.

How many lollies does he buy altogether? 14

4 Year 2 play rounders. They are put into 5 teams with 7 children in each team.

How many children are in Year 2? 35

5 At a railway station there are 9 trains. Eight trains leave the station and 5 more come in.

How many trains are in the station now? 6

6 Every bag has 4 packets of crisps. Luke buys 5 bags.

How many packets of crisps does he buy? 20

7

I am an odd number of 5s less than 40.
I divide exactly by 3.

What am I? 15

Number problems 14

order, round to nearest 10, doubles of multiples of 5, +/– facts to 10

1 Basdev's birthday is on the 14th day of the 7th month of the year.

On what date is Basdev's birthday? 14th July

2 In Year 2, 20 children do not like carrots. Double this number do not like peas.

How many children do not like peas? 40

3 On a farm visit Alex guesses that there are 53 hens.

How many is this to the nearest 10? 50

4 At a barbecue Jenny takes 2 sausages. She later goes back for 7 more.

How many sausages does Jenny have altogether? 9

5 In a small tank there are 35 fish. In a large tank there are double that number of fish.

How many fish are in the large tank? 70

6 There are 8 ponies in the stables. Five of them go into the field.

How many are left in the stables? 3

7

Add 3 to me and subtract 8 from the answer and you get 22.

What am I? 27

Money problems 6

four operations, ×/÷ by 2

1. Leah has saved £2.30 for her holidays.

In the next few weeks she saves another £1.20.
How much has she saved altogether? £3.50

2. Rab has £3.70 birthday money.
He spends £1.10 on a toy car.

How much has he left? £2.60

3. Zack buys two magazines.
Each costs £1.25.

What is the total cost of the two magazines? £2.50

4. Grandma gives her two grandchildren £3 to share equally between them.

How much does each child get? £1.50

5. Fran has £5. She is given another £2. She buys a pair of socks for £1.80.

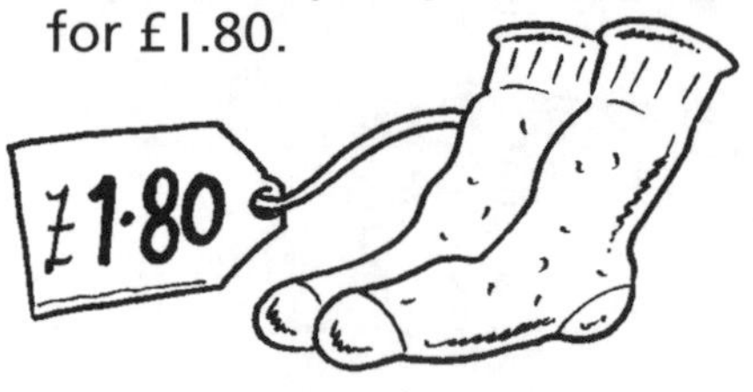

How much has she left? £5.20

6. Cameron has two £5 notes for his birthday.

He spends £6 on a cricket set.
How much has he left? £4

7.

I am the sum of the 2 smallest value notes.

What am I? £15

Number problems 15

facts/pv to ×/÷, 2× facts, pairs of multiples of 10 total 100

1 In a pack there are 8 water bombs.

How many are there in 2 packs? 16

2 There are 300 people in a cinema queue waiting to see Moon Wars. Nine more join them.

How many are waiting now? 309

3 At a sports centre there are 70 children on a list for swimming lessons. Six of them come off the list.

How many are still on the list? 64

4 Liz scores 20 and 80 points on the game board.

How many points has she scored altogether? 100

5 There are 10 skittles and a ball in a box. Shakiel buys 2 boxes.

How many skittles does Shakiel get? 20

6 A train has 65 passengers. At the first station 14 get off.

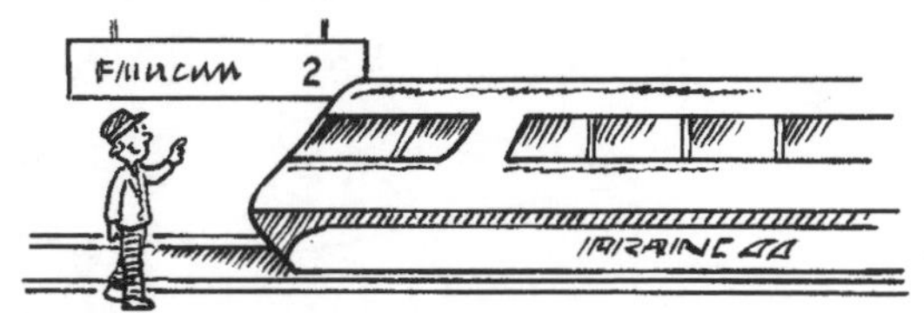

How many are left on the train? 51

7

I am the first number that is in both the 2 times and the 5 times-tables.

What am I? 10

Capacity problems 1

read scales, suggest units, ÷ by 2 facts, 2× and 10× facts

1 A large jug has some juice in it.

Adrian drinks ½ litre of the juice. How much juice is left? 2½ L

2 A barrel of wine holds 16 litres. It is divided equally in to 2 smaller barrels.

How many litres are in each smaller barrel? 8 L

3 Aneesa wants to find out how much water the bath holds.

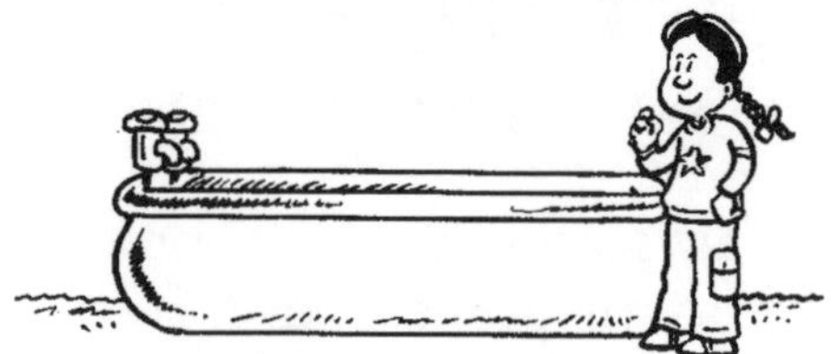

What would be the best unit of measure? litre

4 A bottle of Wizer holds 2 litres.

How much do 10 bottles hold? 20 L

5 Niamh buys twelve 1 litre bottles of milk.

She divides the 12 bottles equally onto 2 shelves. How many litres of milk is on each shelf? 6 L

6 Each pack of coke has 10 one-litre cans.

How many litres are in 9 packs? 90 L

7

I am the second number that is in both the 2 times and the 10 times tables.

What am I? 20

Time problems 4

read half-hours, relationship between units, ÷ by 10 facts

1 The clock at the sports hall says 2 o'clock when Amir and Deepa get in the pool.

They spend 3 hours in the pool.
At what time do they get out of the pool?
5 o'clock or 5:00

2 Ryan flew on holiday at 08:00. The plane was due to leave 3 hours earlier.

At what time was the plane due to leave? 5:00 or 5 o'clock

3 It takes Trudie 1 hour 30 minutes to get to her Grandma's.

How many minutes altogether does it take Trudie? 90 mins

4 Nial's mum looks at the clock when she leaves to go shopping.

She arrive home 2 hours later.
At what time does she arrive home?
1:30 or half-past 1

5 A train arrived at Newcastle station at 1 o'clock in the afternoon. It had left London 4 hours earlier.

At what time did the train leave London? 9 o'clock or 9:00

6 It takes 10 children a total of 80 minutes to do a charity relay run. Each child runs for the same number of minutes.

For how long does each child run? 8 mins

7

I am a digital clock time.
My hours are half of 24.
My minutes are double 24.

What am I?

12:48

Review problems 5

1. How many more children than adults are in the garden? 3
   How many fewer adults than children are in the garden? 3

2. The weights of the boys are in the table.

Don	28 kg
Nick	35 kg
Brendan	27 kg
Brett	32 kg

   Which boy weighs the most? Nick
   Which boy is the third heaviest? Don

3. The Dell family has 2 adults and 3 children.
   They all come late to the barbecue.
   When they come, how many people are there altogether? 20

4. Mr Harris has 3 packs of herb sausages and
   2 packs of spicy sausages to cook.
   How many sausages does he have altogether? 18

5. Mr Watson uses the bucket to fill the paddling pool.
   He puts 10 full buckets of water in the pool.
   How much water is that? 100 L

6. The children guess the total number of crisps There are 60 crisps
   in the large packet and 40 in the small packet.
   How near is Lucia's guess? 10

7. Mrs Dell paid £7 for 2 large packs of burgers.
   Both packs cost the same.
   What was the cost of each pack? £3.50

8. The children had a jumping game.
   Don jumped 1 m 10 cm. Alice jumped twice as far.
   How far did Alice jump? 2 m 20 cm

9. The barbecue started at half-past 5.
   The Harris family left after 2½ hours.
   At what time did they leave? 8 o'clock or 8:00

Number problems 16

count on/back in 3s/4s, +/− 9/11

1 Each pack has 3 pens.

How many pens are in 5 packs? 15

2 Year 2 visit a farm. 64 children and 9 adults go on the visit.

How many people go on the visit altogether? 73

3 Kirstin has 15 plums.
She gives 4 to Irfan and 4 to Nathan.

How many has she left? 7

4 Fran puts 13 biscuits on a plate.

Ryan, Arran and Zahoor each eat 3 biscuits.
How many are left on the plate? 4

5 Video Hire has 87 videos.
Eleven are borrowed.

How many are left? 76

6 Glen has 21 monster cards.
He buys 4 more packs, each with 4 monster cards.

How many monster cards does he have now? 37

7

When you count up seven 3s followed by two 4s you get to me.

What am I? 29

Number problems 17

order, +/– less than 20, + single digit, 5× facts, pairs of multiples of 10 total 100

1 The first house is Number 13.
The last house is Number 21.

What is the number of the middle house? 17

2 An ice-cream man sells 42 cornets and 6 lollies.

How many does he sell altogether? 48

3 Marcus asks 16 friends to his party. Seven cannot come as they are ill.

How many come to his party? 9

4 Each box has 8 muffins.

How many muffins are in 5 boxes? 40

5 Tracy has a box of 8 small dinosaurs and a box of 6 large dinosaurs.

How many dinosaurs has she? 14

6 When playing a computer game Brett scores 10 points and then 90 points.

How many points does he score altogether? 100

7

I am two numbers. The sum of my two numbers is 100. The difference between my two numbers is 30.

What am I?
35, 65

Money problems 7

+/– 19/21,
5× facts,
double multiples of 5

1 Don's dad buys a lettuce and some onions.

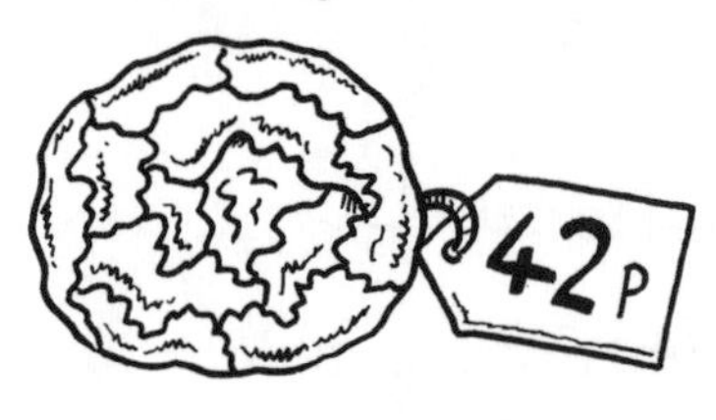

What is the total cost? 61p

2 Each monster costs £5. Grandad buys his 10 grandchildren one each.

What is the total cost of the 10 monsters? £50

3 Tammy takes her purse to the shop. She buys an orange for 21p.

How much has she left? 46p

4 A small box of 10 golf balls costs £40. A large box of 25 golf balls costs twice as much.

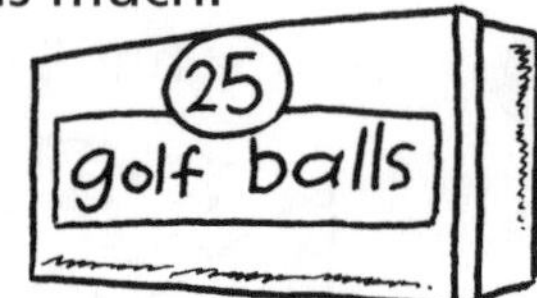

How much is the box of 25 golf balls? £80

5 A shop sells cakes for £6 each. Iain's mum wants 5 of them for a party. She has only got £19.

How much more does she need? £11

6 Rabia shares 14p equally with her brother.
She is then given 21p.

How much has Rabia now? 28p

7

I am 7 of the second smallest value coin.

What am I? 14p

Number problems 18

halving as inverse of doubling, facts/pv to ×/÷, halve multiples of 10, ÷ by 10 facts

1 There are 4 bars of nut chocolate in each pack.

How many bars are in 6 packs? 24

2 There are 30 children in Year 2. Each child has 2 balls. They put all the balls equally into 2 bags.

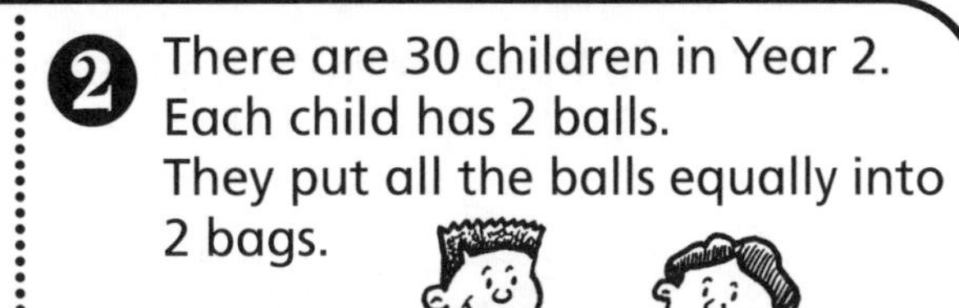

How many balls are in each bag? 30

3 There are 10 eggs in a box.

How many boxes are needed for 80 eggs? 8

4 In a park there are 40 children. Half of them are eating an ice-lolly.

How many children are eating an ice-lolly? 20

5 Holly bakes some cakes.
On each cake she puts 10 jelly sweets.
Altogether she uses 50 jelly sweets.

How many cakes does she bake? 5

6 Meera's mum buys 5 packs of crayons.
She shares them equally between Meera and her 4 friends.

How many crayons does Meera get? 3

7

When you double me and halve the answer you get 50.

What am I? 50

Money problems 8

totals and change, coins, $\frac{2}{4} = \frac{1}{2}$, +/– facts to 10, × by 2

1 Yasmin has 3 coins in her purse. She spends a quarter of her money on a magazine and another quarter of it on a birthday card.

What fraction of her money has she left? ½ How much has she left? £2

2 Pierre buys 2 CDs.
He pays with a £20 note.

What change does he get? £10

3 Carol takes 9p to school.
She loses 4p.

How much has she left? 5p

4 Dreece and Kyle get £3.20 each from their aunt.

How much does their aunt give them altogether? £6.40

5 Callum has £4 in his money box.
He is given another £3.
He spends £1.50 on a book.

How much has he left? £5.50

6 Tess has a £10 note. She buys a T-shirt for £6.
Her dad gives her £2 toward the T-shirt.

How much has she left? £6

7 *I am the sum of half the smallest value note and double the smallest value silver coin.*

What am I?

£2.60

Measures problems 2

read half and quarter hours on clock, pairs that total 20, ÷ by 2 facts

1 A ferry arrives at half-past ten. The journey took 1½ hours.

At what time did the ferry leave?
9 o'clock or 9:00

2 Dina's mum buys 13 litres of petrol for her own car and 7 litres for Dina's dad's car.

How many litres does she buy? 20 L

3 The clock shows the time at which lessons end. Maria leaves school 15 minutes after lessons.

At what time does she leave school?
3:30 or half-past 3

4 A rain barrel holds 18 litres. When it is full Matt pours it equally into 2 smaller buckets.

How much is in each bucket? 9 L

5 At a school fair Sandy sells 16 litres of orange and 4 litres of lemonade.

How much does she sell altogether? 20 L

6 On Monday to Thursday Joe's dad comes home at a quarter to 5. On Friday he comes home ½ an hour earlier.

At what time does he come home on Friday?
quarter-past 4 or 4:15

7

I am between 51 and 79.
I am an even number of 10s.

What am I? 60

Review problems 6

review of previous content

1 Joshua has 75 football cards.
He is given another 11 cards.

How many has he now? 86

2 Toy robots are sold in boxes of 9
Will buys 5 boxes for £14.

How many toy robots does he buy? 45

3 For a party Alexander's mum buys twelve 2-litre bottles of orange juice.
Nine litres are drunk.

How many litres are left? 15 L

4 The clock shows the time Phoebe's bus leaves the bus station.
She has waited 45 minutes for the bus to leave.

At what time did Phoebe get to the bus station?
11:30 or half-past 11

5 On Monday 29 people visited a museum in the morning and 33 in the afternoon. Ten times as many visited the museum on Sunday as on Monday.

Altogether how many visited on Sunday? 620

6 George and Amelia's dad shares £10 equally between them.
George spends £2.50 of his share on a space magazine.

How much has George got left? £2.50

7

When you multiply me by 3 and divide the answer by 4 you get 6.

What am I? 8